TOP-SECRET NATURE

HOW EGGS HATCH

MARIE ROGERS

PowerKiDS press™

New York

Published in 2021 by The Rosen Publishing Group, Inc.
29 East 21st Street, New York, NY 10010

Copyright © 2021 by The Rosen Publishing Group, Inc.

All rights reserved. No part of this book may be reproduced in any form without permission in writing from the publisher, except by a reviewer.

First Edition

Editor: Amanda Vink
Book Design: Rachel Rising

Portions of this work were originally authored by Elena Hobbes and published as *How Do Eggs Hatch?* All new material in this edition authored by Marie Rogers.

Photo Credits: Cover, p. 1 NagyBagoly/ Shutterstock.com; pp. 4,6,8,10,12,14,16,18,20,22 (background) cluckva/Shutterstock.com; p. 5 mayk.75/Shutterstock.com; p. 7 yevgeniy11/Shutterstock.com; p. 9 MR.RAWIN TANPIN/Shutterstock.com; p. 11 kevin leah/Shutterstock.com; p.13 Chokniti Khongchum/Shutterstock.com; p. 15 Jakinnboaz/Shutterstock.com; p. 17 Silarock/Shutterstock.com; p. 19 Anneka/Shutterstock.com; p. 21 Lungkit/Shutterstock.com; p.22 Moonborne/Shutterstock.com.

Library of Congress Cataloging-in-Publication Data

Names: Rogers, Marie, 1990- author.
Title: How eggs hatch / Marie Rogers.
Description: New York : PowerKids Press, [2021]
Identifiers: LCCN 2019044292 | ISBN 9781725317437 (paperback) | ISBN 9781725317451 (paperback) | ISBN 9781725317444 (6 pack) | ISBN 9781725317468 (ebook)
Subjects: LCSH: Eggs-Incubation-Juvenile literature. | Embryology-Juvenile literature. | Chickens-Development-Juvenile literature.
Classification: LCC QL956.5 .R64 2021 | DDC 591.4/68-dc23
LC record available at https://lccn.loc.gov/2019044292

Manufactured in the United States of America

CPSIA Compliance Information: Batch #CSPK20. For Further Information contact Rosen Publishing, New York, New York at 1-800-237-9932.

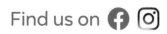

CONTENTS

What's an Egg?	4
Chicks	6
The Nest	8
Egg Needs	10
Turning Eggs	12
What's Inside?	14
Hatch Position	16
Peeps	18
Using an Egg Tooth	20
Welcome to the World!	22
Glossary	23
Index	24
Websites	24

What's an Egg?

Eggs are **amazing**! They can be big or small and many different colors. Some of them even have spots. Many baby animals grow inside eggs. When it's ready to be born, the baby animal **hatches** from the egg.

Chicks

Have you ever seen a chicken? Baby chickens, or chicks, hatch from eggs. This happens about three weeks after a chicken lays an egg. The chick grows in the egg until it's ready to be born. Then it's time to hatch!

The Nest

Chicken eggs have a thin shell and can break easily. Eggs come from female chickens, which are called hens. Hens lay their eggs in a nest made of straw. Some hens lay eggs daily, while others lay eggs once or twice a week.

Egg Needs

Eggs need a lot of care until they hatch. Hens sit on their eggs to keep them safe and warm. Eggs need to stay very warm so the chicks grow. The time during which a chick develops in an egg is called the incubation period.

Turning Eggs

Hens usually lay between 8 and 13 eggs. A group of eggs is called a clutch. Hens turn the eggs to help the chicks grow the right way. They turn each egg many times during the day.

What's Inside?

The inside of a chicken egg includes a clear **liquid** and a **yolk**. The baby chick, called an embryo, uses these as food. At the end of the incubation period, the baby chick is the only thing that fits inside!

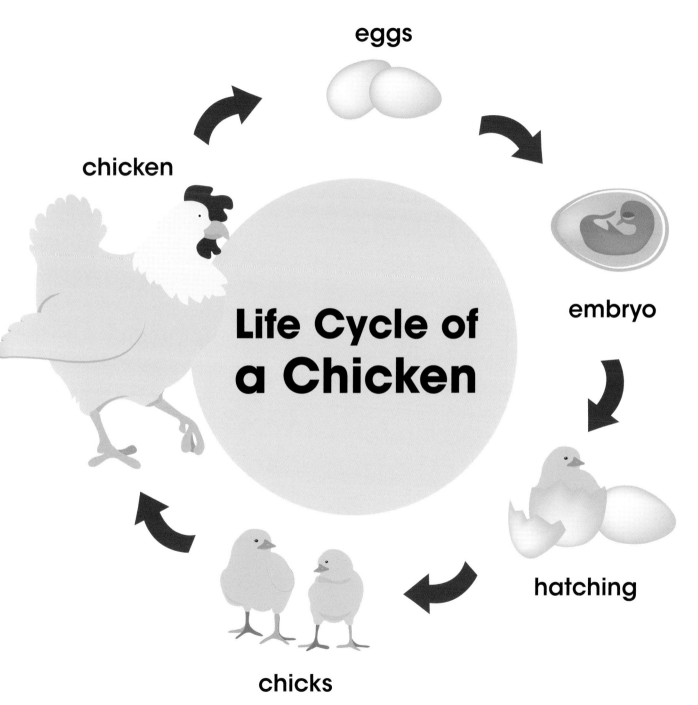

Hatch Position

A few days before a chick hatches, it gets into the right position. Its head is under its right wing. Its **beak** faces the end of the egg that has a small pocket of air.

Peeps

When it's almost time for an egg to hatch, you can start to hear the chick inside! It makes a soft peeping sound from inside the egg. When the hen hears the peeping, she knows her chicks are almost ready to hatch.

Using an Egg Tooth

Chicks have very strong necks. They lift their head up to the shell. Then, they break the shell with their egg tooth. This is a sharp point on the end of their beak. When the chick gets older, the egg tooth falls off.

Welcome to the World!

Finally, the chick breaks through the shell and enters the world. After they hatch, chicks need water, food, and warmth. They have soft, yellow feathers when they are young. They grow up into adult chickens!

GLOSSARY

amazing: Causing great surprise or wonder.

beak: The hard parts that cover a bird's mouth.

hatch: To break open or come out of.

liquid: A substance that flows freely like water.

yolk: The yellow part in the center of an egg.

INDEX

B
baby animals, 4
beak, 16, 20

C
chicks, 6, 10, 12, 14, 15, 16, 18, 20, 22
chickens, 6, 8, 15, 22
clutch, 12

E
embryo, 14, 15

H
hens, 8, 10, 12, 18

I
incubation period, 10, 14

N
nest, 8

S
shell, 8, 20, 22

Y
yolk, 14

WEBSITES

Due to the changing nature of Internet links, PowerKids Press has developed an online list of websites related to the subject of this book. This site is updated regularly. Please use this link to access the list: www.powerkidslinks.com/tsn/eggs